THE LAST AT-BAT

A practical guide to living the
Ranger Creed

By
Stephen M. Davis

This book is dedicated to my wife Judy.

Being on this journey with you by
my side is better than I could
have ever imagined.

INTRODUCTION

The U.S. Army Rangers are the most elite light infantry unit in the world. They are the United States' premier direct-action raid force. For this reason, the Rangers must always stay combat ready. All Rangers have gone through multiple selection and screening processes to make sure only the absolute best make it to a Ranger battalion. If you hear on the news that the United States has sent the Rangers in, you can surmise that we are at war. From before the Revolutionary War to modern times, the Rangers have fought to make sure our freedoms are still intact. Here is a brief history of the Rangers throughout our country's history.

→ French and Indian War – Major Robert Rogers and Captain Benjamin Church both formed Ranger units to fight in the French and Indian War in the mid 1700's. Major Rogers wrote the nineteen standing orders that are still used in the Ranger handbook today.

→ Revolutionary War – Eight companies of highly trained riflemen were formed to fight in the Revolutionary War. These companies were created by orders of the Continental Congress in 1775. Francis Marion, also known as the Swamp Fox, was a consistent thorn in the side to the British attempts to occupy South Carolina.

→ War of 1812 – During the War of 1812, many Ranger companies were organized to protect settlers on the western frontier. From Ohio to western Illinois, mounted Rangers participated in many skirmishes with the British and their Indian allies.

→ The Civil War – John S. Mosby, also known as the Grey Ghost, was highly effective at raiding Union camps and bases. The Northern Virginia area his Mosby's Rangers operated in was known as Mosby's Confederacy.

→ World War II – Six Ranger battalions were activated during World War II. The 1st, 3rd and 4th Ranger Battalions were formed by then Major William O.

Darby and were utilized primarily in North Africa. The 2nd and 5th Ranger Battalions participated in the D-Day landing at Omaha Beach, Normandy on June 6th, 1944. It was here that the Ranger motto "Rangers lead the way!" was born. During the invasion, the 2nd Ranger Battalion scaled the 90-foot cliffs of Point du Hoc to take out German machine gun emplacements that looked out over Omaha Beach. The 6th Ranger Battalion operated mostly in the Philippines and led the raid to free prisoners of war and destroy the POW camp at Cabanatuan. Major General Frank D. Merrill commanded the newly organized 75th Ranger Regiment. The unit became known as Merrill's Marauders due to their effectiveness in the fight against the Japanese during the Burma campaign.

→ Korean War – The Rangers in the Korean War were formed into companies and attached to larger units. 15 Ranger companies were created and were used as internal special operations units for the regiments they were attached to.

→ Vietnam War – Once again the Rangers were called to defend their country and 15 Ranger companies were formed. These Rangers were raised mainly from Long Range Reconnaissance Patrols (LRRP) commonly referred to as "Lurps."

→ Iranian hostage conflict – The modern Ranger battalions were called to action in 1980 when elements of the 1st Ranger Battalion were utilized during the Iran hostage rescue attempt.

→ Operation Urgent Fury – Members of the 1st and 2nd Ranger Battalions participated in a low-level airborne assault in October of 1983. Point Salines Airfield was secured and used as the staging point to rescue American citizens at the True Blue Medical campus.

→ Operation Just Cause – The entire 75th Ranger Regiment was called upon once again for Operation Just Cause in Panama. Simultaneous airborne assaults were conducted at Manuel Noriega's beach house and Rio Hato Airfield to neutralize the Panamanian Defense Forces.

→ Operation Desert Storm – Smaller elements of the 1st Ranger Battalion were deployed to Saudi Arabia in 1991 to support Operation Desert Storm.

→ Somalia – In order to assist the starving and chaos ridden country of Somalia, elements of the Ranger battalions were sent to assist United Nation forces in August of 1993. The Rangers conducted a dangerous daylight raid alongside Special Forces and Detachment-Delta on October 3rd, 1993. In an 18-hour firefight, the Rangers delivered devastating firepower killing an estimated 600 Somalis.

→ Kosovo – In support of Task Force Falcon, the 75th Ranger Regiment deployed a command-and-control element and a Regimental Reconnaissance Detachment in 2000.

→ Global War on Terror – After September 11th, 2001, Rangers were once again called into action to support the Global War on Terrorism. On October 19th, 2001, Rangers from the 3rd Battalion led the ground forces with an airborne assault to seize Objective Rhino in Afghanistan in support of Operation Enduring Freedom. The 3rd Battalion was also called to action to seize Objective Serpent in support of Operation Iraqi Freedom on March 28th, 2003.

Even today, the Ranger battalions are in multiple countries conducting sustained military operations. They continue to stand ready and lead the way when our country calls.

THE RANGER CREED

The Ranger Creed summarizes the ethics, code, and philosophy that all Rangers live by. It creates a connection and loyalty between Rangers of all ages. The creed was written by Command Sergeant Major Neal R. Gentry who was the first Command Sergeant Major of the 1st Ranger Battalion. The Creed is the standard that all Rangers live by when it comes to ethics and how to conduct themselves. As a new private at the Ranger Battalion, I was expected to recite the Ranger Creed word for word on demand. The following pages contain the Ranger Creed in its entirety.

RANGER CREED

Recognizing that I volunteered as a Ranger, fully knowing the hazards of my chosen profession, I will always endeavor to uphold the prestige, honor, and high esprit de corps of the Rangers.

Acknowledging the fact that a Ranger is a more elite soldier, who arrives at the cutting edge of battle by land, sea, or air, I accept the fact that as a Ranger, my country expects me to move further, faster, and fight harder than any other soldier.

Never shall I fail my comrades. I will always keep myself mentally alert, physically strong, and morally straight, and I will shoulder more than my share of the task, whatever it may be, one hundred percent and then some.

Gallantly will I show the world that I am a specially selected and well-trained soldier. My courtesy to superior officers, neatness of dress, and care of equipment shall set the example for others to follow.

Energetically will I meet the enemies of my country. I shall defeat them on the field of battle for I am better trained and will fight with all my might. Surrender is not a Ranger word. I will never leave a fallen comrade to fall into the hands of the enemy and under no circumstances will I ever embarrass my country.

Readily will I display the intestinal fortitude required to fight on to the Ranger objective and complete the mission, though I be the lone survivor.

RANGERS LEAD THE WAY!

PREFACE

This story attempts to connect the lessons that I learned from the Ranger Creed during my time serving in the 2nd Ranger Battalion with the story of a teenager trying out for an elite baseball team. In authoring this book, I thought it would be mostly geared towards the pre-teen/teen age group but have since realized that it has lessons that any age can learn from. The goal is to take the lessons I learned from the Ranger Creed and translate them into a more real world, non-military story. Each chapter begins with a stanza of the Ranger Creed and attempts to impart the lesson from that stanza into the lessons learned while competing with an elite team. My goal for you is to think about your own life and use these principals to potentially raise your own standards and realize your full potential.

Many of the events in the story you are about to read are loosely based on some actual events of my childhood. All the characters in this story are purely fictional. Any resemblance to actual people is purely coincidental.

CHAPTER 1

Recognizing that I volunteered as a Ranger, fully knowing the hazards of my chosen profession, I will always endeavor to uphold the prestige, honor, and high esprit de corps of the Rangers.

To say I had butterflies in my stomach was an understatement. I had finally made it to the tryouts for our town's elite travel baseball team and I was waiting on my turn to take batting practice. I had been dreaming about this moment for as long as I can remember.

I was flashing back to the only other time in my life that I was this nervous going up to bat. It was when I was seven years old and wanted so badly to be a part of the Secret Investigators Club. My brother and a few of the older kids in the neighborhood had created this secret club. They would find clues and solve mysteries around the neighborhood that I grew up in. It was invite only, and according to my seven-year-old brain, very prestigious.

I could not have asked to grow up in a better neighborhood. Marion, Indiana was your typical mid-size midwestern town with the courthouse in the middle of the town square. I grew up on a one-block street that ended in a "T" intersection on both ends. There was no reason to drive down our street unless you were going to your house as there were better routes for travelling that paralleled our street. As much time as possible was spent outside in the summertime. My friends and I would be out right after our bowl of cereal in the morning and barring a quick break for lunch and dinner, could stay outside until the streetlight came on at night. As long as we were within shouting distance from our mothers, we were OK. Those days were filled with bicycle jumps, kick ball games, freeze tag, and kick the can. We knew every good hiding place within a two-block radius.

One of the older neighbor kids had created a secret hideout in his backyard. This was the Secret Investigator Club's top-secret headquarters, and no one was allowed inside who wasn't a member of the club. Our neighbor had an eight-foot fence around their backyard that came together in a point at the very back of the property. There they had connected a line and hung a big blue tarp to block out any intruders. As I came to find out later, each member had their own mat to sit on, and under one of the mats, all the clues were buried in a coffee can. These were the clues to whatever cases they were solving at the time. Under no circumstances could you tell a soul where the clues were hidden. At the time, this was the most important day of my life. My chance to be part of this elite club.

The tryout for the Secret Investigator Club had nothing to do with how good you could solve a mystery. Nothing to do with your critical thinking skills or attention to detail. The prospective candidate had to hit a baseball past the light pole that was a few houses down the street from where we lived. This was the same light pole that told us when we had to be in for the night on many summer evenings. The ball had to travel just one hundred feet or so, but it was the most important at-bat of my life up to that point.

I had three pitches to hit the ball in the air past the light pole or I could not be a part of the club. With my older brother pitching, I hit a groundball on the first pitch. Disappointed, I got back up to home plate, which was a filled-in pothole in the middle of the street. This pothole was used for our home base many times for a baseball or kickball game in the street. On the second pitch, I lofted the ball into the air, but it was just not long enough to clear the light pole. This was it. I had one more at-bat to make my childhood dream come true.

On the third pitch, I connected well, and it cleared the pole by a good 20 feet. I felt like Babe Ruth! I jumped for joy and got several slaps on the back by the other members of the club. It felt so good to be a part of such a prestigious and private club.

Now that I was a part of the Secret Investigators Club, we all carried a pocket-sized notepad on which we wrote down clues. We would put a normal list in the front of the notepad, but if you flipped it over and started from the back, you would find the various clues of the cases we were working on. My seven-year-old brain thought that was so clever. If one of the neighbors was missing something, not to be worried. The Secret Investigators were on the case!

"Jones!" I snapped back to the present moment. "You're on deck!" Okay, this is it. My mind raced back to reality and the gravity of the situation. It was eight years later, and I wasn't trying out for the Secret Investigators Club. This was the tryout for the Grant County Bombers!

The Bombers were the most elite travel baseball team in our community. If you were a kid growing up in Grant County, Indiana and played baseball, you had a dream of playing for the Bombers one day. When I was a kid, they were the stuff of legend. I remember watching them when I was younger and was amazed at how good they were at every aspect of the game of baseball. They had won tournaments for years all over the state and were the automatic favorites in any tournament they entered. They had won the Midwestern Fall Classic the last three years in a row. The Midwestern Fall Classic was a tournament in September that brought the best teams from six of the surrounding states.

The tryouts for the Bombers were demanding and the expectations were high. There were around fifty kids trying out that day and only fifteen players would make the team. Almost half of the kids trying out were from my school. This was day one of a two-day tryout, and to-day's focus was hitting. I had worked my way up through tee ball and little league and performed well. I played second base my freshman year for our middle school team. I was always a little scrawny but had a lot of drive and was a consistent hitter. Not much for power but bat-ted sixth with a .311 batting average my freshman year. The competition would be fierce, but I knew if I gave it everything, there was a chance I could make the team.

"Okay, let's see what you got, Jones!" yelled the assis-tant coach from the pitcher's mound. He was behind a pitcher's L screen with a five-gallon bucket full of base-balls. It was all a blur, but he threw me probably 25 pitch-es, all fastballs. I connected on around 20 and had proba-bly 11 or 12 good rips to the outfield. I walked away encouraged and thought I had done a decent job. I hoped it was enough to get the coaches' attention because I had never wanted anything more in my life.

In between hitting we shagged balls in the outfield. I could tell some of the kids did not belong and were not taking things seriously. I moved up to second base to hopefully show some of my defensive skills. After two additional rounds of hitting, day one practice was over. I felt confident that I was one of the better hitters that day. My second and third rounds of hitting went much like my first and I left excited for the next day's tryouts.

Day two was much more physically demanding. We started out running wind sprints. The coaches had us line up behind the third base line facing right field. On the command of "Go!" we would sprint to an imaginary line between first and second, back to the third base line, sprint to and touch the right field wall, and then run back to the third base line. The purpose of this was not only to see who was physically fit, but more to get us mentally and physically exhausted to see how we would perform the rest of the practice. This simulated the eighth or ninth inning of a game when we would be mentally and physically spent.

I noticed they had a coach stationed at each line where we turned around. Some of the kids would turn just a little shy of the line or not quite touch the outfield wall. These kids thought that they were getting an advantage, but in hindsight I realized none of those kids made the team. Years later when speaking with one of the coaches, he said they would look for players who would not go the full distance or touch the line. He told me these players were never selected as they would let down their teammates when they got under some pressure or when things got tough. Fortunately, I made sure my foot always went over the line and that I always touched the outfield wall. We ran wind sprints until some of the players were cramping or felt like they were going to throw up.

After that we were put into every defensive situation possible. We were hit flyballs and grounders and put in double play, steal, and bunt situations. This was to test our baseball IQ and to see how we would perform after being run half to death. Luckily, I had some good plays in the field. There was only one player at each position at a time, so the other players would put on batting helmets and run the bases. The coaches would put players on base and hit the ball to a certain spot to see if we knew where to go with the ball. I turned a few double plays at second base. Some kids would bobble a grounder and let it affect them for the next few plays. You could see them with their heads down and negative body language. I noticed a couple of the coaches pointing this out and I made sure I kept my head up and was always encouraging to my teammates.

The coaches finally called the practice and I jumped on my bike to go home. The results of who made the team would be posted on the team website by 6 p.m. the next evening. I couldn't even eat dinner that night I was so nervous. I lay in bed thinking of how awesome it would be to be a part of a great team with such a winning track record. I had given my all and left everything on the field. Make the team or not, I was proud of the effort I had given. Somehow, I drifted off to sleep dreaming of hitting balls into the gap between center and right and stretching it for a headfirst slide safe at third.

The next day I hung out with my best friend, Kevin Price. Kevin lived a couple of neighborhoods away, but it was only a 10-minute bike ride to my house. Kevin and I had been best friends since we played tee ball together when we were eight years old. Kem Road Standard was our team. I remember the night I got the call from our coach to let me know that I was playing on that team.

Kevin was the shortstop and I played second base. We had a lot of fun that summer trying to turn double plays. A purple T-shirt with Kem Road Standard on the front and a pair of blue jeans was our uniform.

We spent many summer days at the Police Athletic League ball field. We then moved on to Lincoln Field, where we played for Kiwanis in Little League. Kevin always at shortstop and me at second base. Kevin was a couple of inches taller and a little stockier than me. This made him a better power hitter. Kevin had blond hair, which was usually covered up with a ball cap, and blue eyes. I had noticed he was starting to get attention from some of the girls at school. He got some baseball recognition as well at the little league level, making the all-star team his last two years playing little league. I joined him my last year of little league and we got to start in the middle infield together, representing our team for that game. This last year we played for our high school freshman team. We started every game together. Kevin batted fifth and batted .347 his freshman year. He was on track to do great things, and with the right dedication, had the potential to play at the college or even the professional level. Kevin was a great friend and teammate. He was always encouraging even when you struck out or made a bonehead play.

"Do you think we made it?" I asked Kevin as we played catch in my front yard. "Of course we did!" he said without missing a beat. Once again, he was upbeat and positive. I was more nervous than ever. It was great having Kevin there so that I didn't have to think too much about it. We spent a summer day like most kids do—whatever grabbed our attention next. It could have been jumping ramps with our bikes, playing kickball or baseball in our street, or trying our best to awkwardly flirt with some girls in our neighborhood.

I was at the computer at 5:45 p.m. I had never hit the "refresh" button so many times in my life. I had never wanted something so badly, and my stomach was telling me so. At 6:01 p.m., after hitting refresh at least 200 times, the screen changed, and the results were posted. I quickly scanned down the row. Cooney, Diaz, Donnelly, Gannon, Golden, JONES! There it was! I had made the team! I quickly scanned down the list to see if Kevin had made it as well. PRICE! Kevin had made it too.

I was a Grant County Bomber! I felt so honored to now be in the same company as so many of the talented players that were on the team before me. For the rest of my life, I could say I was a part of this prestigious team. Little did I know that the hard work and sacrifice was about to begin.

LESSONS FROM THE CREED:

If you are going to be a part of an elite team, expect to work harder than you ever have to earn the right to be there. You should recognize the price the players who went before you had to pay and honor their commitment to excellence. You should always have the attitude that you will not let your team members, friends, or family down.

CHAPTER 2

Acknowledging the fact that a Ranger is a more elite soldier who arrives at the cutting edge of battle by land, sea, or air, I accept the fact that as a Ranger my country expects me to move further, faster, and fight harder than any other Soldier.

Practices were from 5:30 p.m. to 7:30 p.m. on Mondays and Wednesdays and on Fridays when we didn't have a tournament that weekend. Practices were demanding and the expectations were that you would not miss any. Kevin and I showed up together to our first practice. We arrived 30 minutes early and still didn't beat the coaching staff there. We just started warming up, tossing the ball together.

Brian Bowman was our head coach. He wore a high and tight haircut and looked like he still spent time in the weight room. He had a square jaw and looked like he could have substituted as an Army drill sergeant on the weekends. He was tough. He was demanding. He expected the absolute best out of you, and if he didn't think he was getting it, he would definitely let you know. Coach Bowman had been a catcher in his younger days and even played in college. He rode the bench most of his freshman year at Coastal Carolina University. The starting catcher was a senior, and Coach Bowman was patient and learned as much as he could from him. As a sophomore he became the starting catcher. Major League teams were scouting him, and many thought he would play at that level. It was only when he blew out his left knee during a slide into third base stretching a double into a triple that he knew his baseball career was over.

Coach Bowman's grandfather had been a marine in World War II. He was part of the 2nd Marine Division and part of the fleet that landed on Okinawa. He was wounded by shrapnel that went into his left arm and received a Purple Heart for being wounded in combat. One interesting fact that not many people knew was that Coach Bowman's grandfather didn't get drafted or didn't have to go fight in the war. He had two young children and was exempt from serving. Coach Bowman's father was just a baby at the time. Only when he heard the news of one of his best friends being killed in France did he go down to the recruiting station and enlist.

Coach Bowman remembered countless stories that his grandfather told him about basic training and the drill sergeants. One story he told was a soldier who had dropped his weapon in the rain while training. He had to run around a large quad area with his weapon held over his head. The entire time it was pouring, and they made him run for a couple of hours. I could see that Coach Bowman's discipline coaching style was influenced by lessons from his grandfather. Coach Bowman has some letters that his grandfather wrote his grandmother during boot camp and while he was in the hospital. In these letters, he is very loving to Coach Bowman's grandmother and refers to Coach Bowman's father as "little Jimmy." Coach Bowman's grandfather passed away several years ago, and one of Coach Bowman's most prized possessions is his grandfather's Purple Heart that he received for being wounded in combat. This is the kind of person Coach Bowman's grandfather was, and you could tell that same sense of honor and courage was instilled in Coach Bowman.

I will never forget that first practice. With a shout of "huddle up!" Coach Bowman motioned for everyone to come toward him. After introducing us to the coaches, he told us exactly what we could expect from our practices and from him. He let us know that the practices were going to be tough and demanding both physically and mentally. Then he proceeded to inform us that his expectations were going to be equally demanding. He was going to give it his all and he expected the same out of us.

We started off with a slow run jogging around the outfield fence to get loose. This was not a race as the sole purpose was to warm up. When we got back in, we gathered in the outfield grass just behind second base. The fifteen of us sat around one of the coaches in a circle. The coach led stretching for a full 10 minutes, and we mirrored the stretches as he was doing them.

We then warmed up by throwing. Starting in pairs, we began about fifteen feet apart and on one knee. The coaches constantly corrected our form and follow-through. After this, we stood stationary from about the same distance. Only after several minutes did we do a full throw and step into it. You would think this was just a warm-up, but the coaches gave me some pointers and I could really tell the difference. I was using the same energy but had more power and accuracy.

It was then time for groundballs and flyballs. We divided into two groups. One group took turns at shortstop fielding groundballs and the other group went to left field. There, one of the coaches would hit us flyballs from around the right field foul line. Another coach functioned as catcher and received the throws back from us. The coaches worked with us on our footwork and throwing back to the catcher. After about 10 minutes, we switched, and I was now fielding groundballs. The coaching was amazing. We worked on keeping our momentum toward the base we were throwing to and staying low during the throw. I knew I was learning at a whole new level now, and my game would improve dramatically because of it.

We spent the next half hour or so on team fundamentals. First, we worked on bunt defense. I gave Kevin a quick grin as he ran past me on his way to running his first lap. We were simulating a runner on first and second and the batter laying down a bunt. Kevin's first instinct was to cover second base, but on this drill, he was supposed to cover third. The coaches expected excellence, and we heard the coaches yell the familiar, "Run a lap!"

We then worked on runners on first and third defense, cutoffs and relays, pickoffs, rundowns, and we finally did some baserunning practice.

"Jones! Run a lap!" Unfortunately, this was heard more than once during practice. Coach Bowman demanded excellence. This time it was for not positioning myself correctly as the cutoff man for a throw from the gap between right and center field. If I positioned myself a little deeper, the relay throw would have gotten to home that much quicker. This is the knowledge and attention to detail that made the Bombers a championship team. Luckily, Coach Bowman did not discriminate on who he handed out laps to. All fifteen of us usually had to run at least one lap during most practices. Coach Bowman never made you run a lap just because he was the coach. He had high standards and expected you to know your stuff. He was fair.

Practice was well planned out and ran like a machine. After a predetermined time, one drill would end, a whistle would blow, and we would be on to the next.

The last 45 minutes or so was batting practice. We all got four or five minutes of hitting, with the coaches working on our fundamentals.

We huddled up at the end of practice. "I liked what I saw out there today," Coach Bowman said. His voice was calm but still had a deep, stern tone to it. "There are fundamentals that need to be worked on, but we can take care of that with repetition and practice. That can all be taught. What I liked best was the effort that I saw each of you putting into every drill. Your effort, attitude, and encouragement of your teammates will win you more championships than raw talent ever will."

We all listened intently and ate up every word. Coach Bowman had the hardware to prove that he knew what he was talking about. I left that first practice feeling tired but great. I knew I would have to give my best every practice, but it would all be worth it. My teammates and coaches were counting on me.

LESSONS FROM THE CREED:

If you are going to be a part of an elite team, the expectations and demands on you are going to be much higher. It will be one of the most challenging things you have ever done. You will have to stretch and become a better version of yourself, but it will all be worth it when you earn the right to be a part of something so prestigious.

CHAPTER 3

Never shall I fail my comrades. I will always keep myself mentally alert, physically strong, and morally straight and I will shoulder more than my share of the task whatever it may be, one hundred percent and then some.

The first time Kevin and I saw them was about 30 minutes before one of our Friday practices. We were in our third week, and we saw the same group of kids hanging around and talking with a few of the players. I recognized one of the kids from my school. His name was Joe Stevens and he was the kind of kid who didn't care much about school. Joe had long brownish hair that usually hung in his eyes and looked like it hadn't been washed in a few days. He constantly was late to class and spent most of his time in detention. He was wearing a green Army jacket and was hanging out with a couple of other kids I did not recognize. There was a cloud of smoke above them as they all were smoking cigarettes.

"Hey, what's up?" Joe said, as Kevin and I were walking over to practice. We acknowledged them with a "Hey" but didn't say much more than that as we continued to walk toward the field. "Hey, come here for a minute," he said to us. Kevin and I walked over a little cautious to see what they wanted.

"Aren't you in my math class at school?" Joe said.

I said, "Yeah, I think so." Even though he was couple of years older than us, Joe was at my math level.

"So, you guys are Grant County Bombers, huh?" he said with a smirk on his face.

"Damn right we are!" I said with conviction. I could feel my adrenaline pumping a little.

Joe said, "That's cool, man. I wasn't making fun of it. As a matter of fact, I am good friends with a couple of the high school players that used to play on the Bombers. I would even go as far as to say I am the reason that they made the high school team in the first place."

Kevin and I looked at each other and rolled our eyes. "Yeah right," Kevin said. "Dude, what are you even talking about?"

At this point Joe threw his cigarette down and snuffed it out with his shoe. He looked around to his right and left and reached into his pocket. He pulled out a little plastic bag with some red and blue pills in it. "These will make you play like Babe Ruth!" Joe said with a smile on his face.

Kevin and I were old enough to remember the scandal in the Major Leagues concerning performance-enhancing drugs or PEDs. Players had set MLB home run records but would never make the Hall of Fame because they cheated by using PEDs. Their reputation would be forever tarnished, and their families are still shamed to this day. There was no way that we were going to jeopardize our reputation and embarrass our families and our team like that. Kevin and I could not believe that we were being offered these at the travel baseball level. We would be immediately kicked off the team and likely never be able to play again. There was no temptation at all.

"Man, get the heck out of here!" I yelled at Joe. Joe shoved the bag back into his pocket and said, "It's cool, man."

I said, "No, it is NOT cool! You are putting us and every one of our teammates in danger, and it could ruin any chance we have to play at the next level."

Kevin chimed in and said, "As a matter of fact, if we ever see you around here again talking to any of our teammates, we are going to tell the coaching staff and I hope they call the cops on you!"

Joe just shook his head and nodded at his buddies. "Let's roll," he said to them, and he pedaled off toward the exit road. That was the last time we ever saw Joe around the baseball field.

It took us a few minutes, but Kevin and I were able to get our heads cleared and focused on having a great practice.

This was the third week of practice, and we could see the team starting to get what the coaches were teaching. We were clicking as a team, and everyone encouraged each other. The other good news was that we didn't hear "Take a Lap!" as much from Coach Bowman as we were starting to put into practice what the coaches were teaching. At least that is what I thought until this practice.

I don't know if it was the kids offering us drugs or something else, but Kevin had a brain fart during the second hour of our practice. The coaches put one player at each position on the field and the other teammates were running the bases. The six players not fielding wore batting helmets, and the coaches would put them at different bases simulating different gametime scenarios. We would practice fielding with zero, one, or two outs to make sure we knew exactly where to go with the ball if it came to us.

Kevin was one of the runners and had his batting helmet on standing by home plate. This was just a simple groundball to the infield and throw the runner out at first. "Get one!" Coach Bowman yelled as he hit a groundball to shortstop. This was normally Kevin's position, but since he was a baserunner, Alan Martinez was filling in for him. Alan was a decent infielder and could play second, short, or third. A switch-hitter, he was a great player to use as a pinch-hitter late in a game.

Alan fielded the groundball cleanly in between the second and third hop. His footwork was perfect as he rounded, set, and then stayed low on his throw to first.

Kevin took off like a sprinter at the crack of the bat. He wasn't the fastest kid on the team but had decent speed for someone his size. He saw Alan field the ball cleanly and make a good throw to first. For some reason, about fifteen feet from first base, Kevin let up and jogged the rest of the way to the bag. This was never tolerated by our coaching staff. You were expected to run as hard as you could, hitting the front of the bag and running through the base. What made matters worse was that Billy, our first baseman, bobbled the throw. The ball actually dropped to the ground, and he was able to reach down and pick it up before Kevin made it to the bag. Kevin would have been safe if he kept running hard all the way through the bag. Kevin ran past the bag and immediately put his hands on the side of his helmet. He knew he had messed up. Coach Bowman stopped practice and called everyone to the pitcher's mound. We knew it was not going to be pretty.

"This is how you lose baseball games," Coach Bowman said. "This is how you let your teammates down." Kevin just looked at the coach. He knew not to hang his head as this would make matters worse. "Your teammates will always know when you are slacking, when you are not giving it your all. This will not be tolerated on my team and is the fastest way to ride the bench. Kevin, you know what to do." Kevin took off to start running a lap. "Don't stop until I tell you to!" Coach Bowman yelled behind him.

I never felt so bad for my best friend. Normally a player would run a lap and come right back to rejoin practice. Coach Bowman made Kevin run for the next 30 minutes. I could tell he was smoked. He only stopped when it was time for him to take batting practice. Coach Bowman was obviously driving the point home. I tried to encourage Kevin during batting practice and would yell out "Good rip!" when he hit a solid drive. I could tell his world was rocked. The last thing Kevin wanted to do was to let down one of the coaches or one of his teammates.

We finished practice and were packing up our gear. We were getting ready to walk over to our bikes when Coach Bowman yelled, "Price! Come here!" I watched as Coach Bowman put his arm around Kevin and they walked down toward third base. They talked for a good five minutes then Kevin came back with a little more spring in his step.

On the way home I said, "You don't have to tell me if you don't want to, but what did Coach Bowman say to you?" Kevin said that Coach had made an example of him for a reason. He knew Kevin was one of the leaders on the team, and that he knew he would take the constructive criticism the right way. He knew that the other players looked up to Kevin and that this would drive the point home. Not giving one hundred percent all the time was not acceptable on the Bombers. Coach Bowman knew that the other players on the team would get the message and that Kevin was the one to set the example. He let Kevin know that he expected he would never let up again and that his teammates could count on him. Kevin promised himself that day that he would give one hundred percent, and then some, for the rest of the season.

LESSONS FROM THE CREED:

Your friends and teammates are counting on you. They will know if you are not giving everything but your absolute best. Live a clean life and do not compromise. Do not put anything in your mind or body that will cause it harm. Give everything you do everything you have. If it is worth doing, it is worth doing right and giving it your all.

CHAPTER 4

Gallantly will I show the world that I am a specially selected and well-trained Soldier. My courtesy to superior officers, neatness of dress, and care of equipment shall set the example for others to follow.

After six weeks of practices, we were entering our first tournament. Our team was firing on all cylinders. We were not cocky, but we definitely had some swag. Knowing we had paid the price and practiced hard, we were going to be a tough team to beat in any tournament. As usual, Kevin and I got there about 30 minutes before the coaches wanted us to be there. We were usually the first two players to arrive. Only the coaches got there before us.

"Martinez, tuck your shirt in!" yelled Coach Bowman. Martinez had just arrived and was walking toward us.

Coach Bowman didn't tolerate any players wearing an untucked shirt or having their hat on backwards or sideways. Coach always said how you look in your uniform says a lot about who you are as a person. "Do you have pride in yourself?" Coach Bowman asked us all at practice one day. We all answered with a "Yes, coach!" He then went on to say "How you dress, how you take care of your gear, this will speak volumes. People will judge you by your actions and not your words or your intentions." These words still ring true to me today.

"Yes, Coach!" Martinez yelled as he tucked in his shirt.

Coach Bowman didn't give a crap about what other coaches let their players do. We played teams that had players wearing different styles of pants and tucked and untucked shirts. Not Coach Bowman. He believed that if we were a team, we would all wear our uniforms the same as a sign of unity and solidarity. Uniforms were to be clean, and shirts tucked in at all times. He wanted us to represent excellence.

We won our first two games on Saturday easily. The first game we won 14-2 and the second we won 9-4. I went only 1-for-4 in the first game but made a great defensive play. It happened in the top of the sixth. I robbed a base hit up the middle with a diving stop to my right. I was then able to flip the ball to Kevin at second base for the force-out. I went 3-for-5 in the second game with a double and two singles. We were now 2-0 after our Bombers debut. This would put us in the championship game on Sunday.

Our offense was tough to pitch around. We really didn't have a weak spot in our lineup. Rick Diaz played center field and led off. He was the fastest kid on our team and could really fly down the line. I saw him beat out throws to first on a bunt, and occasionally, he would beat out a groundball. Jimmy Gannon played third base and batted second. Jimmy was one of the better hitters on our team. He didn't hit for a lot of power but had a great eye and was always a tough out for the other team. Even though Diaz was faster, Gannon led the team with the highest on-base percentage.

Next came the power hitters. It was Diaz and Gannon's job to get on base and it was the next three batters' job to knock them in. Bill Cooney played first base and batted third, and Roger Metzler played left field and batted cleanup. Kevin batted fifth and was great protection for Roger. There was no way a team was going to pitch around Roger to get to Kevin. I batted sixth right after Kevin. Schreiber, Williams, and Donnelly, who was pitching for the championship, rounded out our lineup.

We were tough to beat with that lineup. We practiced just as hard as we played in the game. The coaches were always there to make slight corrections and offer words of encouragement. What was great to see early on is that we had the same expectations of each other. We knew that not only our coaches were watching, but our teammates as well, and we were not going to let each other down.

I had butterflies in my stomach on Sunday before the game. This was my first championship game as a Bomber! We were the home team and were lined up along the third base line for the National Anthem. Like always, Coach Bowman was a stickler for unity. We stood facing the American flag, hats off and our hands over our hearts. Coach Bowman wouldn't tolerate anyone talking during the Anthem or putting their hands behind their back. He could care less what the other teams did. He appreciated what the flag stood for and always said that many men had given their lives for us to even have the freedom to play baseball.

The team we were playing would be our biggest challenge yet. They were from Clark County, Indiana and were called the Cougars. James Churry was the manager of the Cougars and he and Coach Bowman had battled for several years now. Coach Churry had played baseball in the minor leagues. He was a standout at Indiana University and then played in the Milwaukee Brewers' minor league system. A right-handed pitcher, Coach Curry made it to Double-A before he realized that he just wasn't going to make it to the big leagues.

Coach Bowman had nothing but good things to say about Coach Curry. He shared many of the same values such as working hard and setting high standards. They chatted a little bit behind home plate before the game. They shook hands at the end of the conversation and Coach Curry said, "Good luck!" Coach Bowman smiled and said, "May the best team win." You could tell they had a lot of mutual respect.

We got on the scoreboard in the top of the first inning. Diaz ripped a leadoff base hit right over the shortstop's head. The shortstop leaped and the ball cleared his glove by about six inches. Gannon then laid a perfect bunt down the third base line to advance Diaz to second. Cooney struck out on a wicked curveball. He took a good cut at it but was way out in front. Metzler got us a run on a line drive down the left field line. Even though Metzler was thrown out trying to stretch it into a double, Diaz easily made it home with his speed to put us up 1-0.

The score stayed that way until the bottom of the fourth. The Cougars put together a walk, a double, and then three base hits in a row and were now up 3-1. This was the first time we had ever even been behind in a game. We were going to see how we managed adversity. There was no quit in the Bombers! We had the talent and did the work. Never once did I even think that we would lose the game.

Jason Adams was the Cougars' pitcher. He had a fastball in the mid-seventies that was tough to catch up with. Combine that with a nasty curve and change-up and he could make you look silly. There were rumors that some college scouts had already seen Jason pitch and that he was probably going to get a full ride some-where.

Jason had settled in after the first inning and had a two-hitter going through five innings when we came up in the top of the sixth. I would bat fourth this inning and was already 0-for-2 with a strikeout and a groundout to the second baseman. We had Cooney, Metzler, and Price coming up, the heart of our lineup. We were all expect-ing to put some runs on the scoreboard. Unfortunately, Cooney struck out again. He was not having a good day; he was now 0-for-3 with three strikeouts. This was Ad-ams' eighth strikeout of the game with no walks. He was living up to his reputation as a great pitcher.

Metzler got us going with his second hit of the day. It was a hard groundball perfectly placed between first and second. Kevin was up next, and we all encouraged him to keep it going. He quickly got down 1-2 and we knew Adams was not going to throw him something to hit. Sure enough, he threw a nasty change-up that dropped into the dirt about a foot before home plate. Kevin couldn't resist what looked like a fat pitch coming right down the middle and swung away. Somehow, Kevin caught a piece of the ball and it dribbled down the third base line. It was like a bunt and was hugging the line, not sure if it wanted to be foul or fair. The third baseman, playing deep, charged the ball hard. Adams went running over to field it as well. They both got to the ball around the same time. All it took was a moment of hesitation to decide if it was going to be fair or foul and who was going to field it to make the difference. It was like time froze as they all stood and looked at the ball. It finally rested just an inch inside the foul line. It was a fair ball! Now we had something going! I was up with a man on first and second and only one out.

In my mind, I was just telling myself to be patient and wait for a fastball that I could get ahold of. Adams threw a change-up with the first pitch. I swung over the top of it by about a foot. He made me look silly. The second pitch was a fastball but was a little low and outside. I was able to hold off and the count was 1 and 1. The third pitch was exactly what I was looking for. A fastball right down the middle. I swung and made good contact. I thought I had a base hit between third and short, but their shortstop made an incredible stop. Going to his right, he backhanded the ball right in front of the out-field grass and popped up firing to second base. I took off down the line and ran as hard as I could. There was no way I was going to let them turn two. Kevin slid into second base but was out by a couple of feet. He tried to break up the throw to first with his slide, but the second baseman made a textbook relay.

Pop! Pop! I heard in quick succession right after each other. Fortunately for me, the first *pop* was my foot hitting the bag and the second was the *pop* of the ball hitting the first baseman's glove.

"You're out!" I heard the first base umpire yell as he clenched his fist in the air. I couldn't believe it. I immediately looked at Coach Bowman and he was already coming out of the dugout. I started to walk over with him, and he pointed to the dugout and said "Jones, get in the dugout!"

Coach Bowman was old school and played the game the right way. He taught us to never disrespect the umpires, other coaches, or the other players. I could tell he was hot. Coach Bowman was not a "throw my hat or kick dirt on the umpire's shoe" type of arguer, but he let the ump know that he had blown it. With no instant replay, he knew the call would not be reversed but appealed to the other umpires anyway. Out of respect, they all said that the ump who had called me out had the angle and was in the best position to make the call.

Coach Bowman argued for a few more minutes and then walked back to the dugout. We knew he had our backs and wouldn't stand for a blown call. The call stood and we were out of another inning still down 3-1. Coach Bowman pulled us together quickly before we took the field. "Everyone, keep your head up! That call is in the past and we can't do anything about it now. Let's forget about it and keep playing hard. I want to see everyone giving one hundred percent. Let's go!" We ran out of the dugout fired up and ready to redeem ourselves.

I don't know if it was the bad call, the argument with the ump, or Coach Bowman's pep talk, but we came alive in the last three innings of the game. We scored a run in the seventh and three runs in the eighth and tacked on another one in the top of the ninth. The final score was 6-4, and we had our first championship. Cooney redeemed himself in the top of the eighth. With the bases loaded, he cracked a double to the gap between left and center that rolled all the way to the fence. The throw to home was wide and allowed Diaz to score all the way from first base. Cooney pumped his fist as we all yelled from the dugout. Mark Golden, our closer, came in to pitch the bottom of the ninth and we had a three-up, three-out inning for the win.

We lined up to shake the other team's hands. Coach Bowman tolerated no trash talk or cockiness. You said either "good game" or could say something positive about a good play or at-bat a member of the opposite team had. We always kept it positive. You never rubbed a win in their face or bragged. We then got back to the dugout and high-fived and hugged each other. It felt good to see our team come from behind for the win. We got behind in runs and had a couple of bad breaks, but we still stuck together as a team, encouraged each other, and came away with a victory. It felt good to be a Grant County Bomber, and it felt even better to win my first championship with our team!

LESSONS FROM THE CREED:

If you are going to be on an elite team, then every action you take should set an example. How you treat others, yourself, and the things you own speaks volumes. How you practice determines how you play when it is game time. Your colleagues and teammates are watching to see if you respect your coaches, yourself, and the things you own.

CHAPTER 5

Energetically will I meet the enemies of my country. I shall defeat them on the field of battle for I am better trained and will fight with all my might. Surrender is not a Ranger word. I will never leave a fallen comrade to fall into the hands of the enemy and under no circumstances will I ever embarrass my country.

It was one of the best summers of my life. The comradery I built with my teammates is still as strong as ever, even today. Many, including Kevin, are still some of my best friends. We entered five tournaments around the state that summer and came away with five championships. We still practiced hard every week and the coaches never compromised on their high standards. This made us even more confident that we could beat any team, and we went into every tournament expecting to win. We fell behind only three times during those five tournaments. Even then, it was only for an inning or two and was never more than two runs. Even when we got behind, we never thought we would lose. In fact, it got us fired up! We seemed to encourage our teammates and increase our efforts even more. The coaches weren't the only ones with high expectations. We had high expectations of each other and didn't allow any of our teammates to perform at a level that we didn't think they were capable of. Sometimes I felt I didn't want to let my teammates down even more than the coaches.

All summer long we kept hearing about the Jefferson County Cardinals. Even though they were two states away, we always knew how they were doing and if they had lost any games or tournaments. They also had a rich history of winning and were by far the best team in their state. Our coaches never even mentioned them, but we always heard through a friend or someone at school how they did in their tournaments. You see, either the Bombers or the Cardinals had won the last 11 out of 12 Midwestern Fall Classics. Six of those belonged to us and they had the other five.

Sure enough, they too had won every tournament they entered that year, and they were coming to Indianapolis to meet us in the Midwestern Classic. This year the tournament would be held at Victory Field, home of the Indianapolis Indians. The Indians are the Triple-A affiliate of the Pittsburgh Pirates and have played there since the stadium was built. The stadium could hold over 12,000 people and was the site of the Indiana state high school baseball championships each year.

We were ready! I think we were actually excited that the Cardinals were going to be there and would be gunning for us. To be the best, you must beat the best, and we were eager for the challenge.

Frank Dowd coached the Cardinals. He had been a standout at the University of Missouri. He played right field but was more known for his power. His achievement of 37 home runs in 58 games is still the single season record at the University of Missouri. He had a great four years there and was a second-team All-American his senior year. He did get a lot of attention from the Major Leagues and had scouts present at most of his games during his senior year. He was drafted in the third round, 96th overall, by the Detroit Tigers. He fast-tracked through Single-A and Double-A and spent three years in

Triple-A before he was finally called up at the age of twenty-five.

Dowd played a half season with the big-league team before he was sent down. Unfortunately, the opposing pitchers figured out how to pitch Dowd pretty quickly. He hit only .167 with four home runs and 13 RBIs in 78 games that year. The coaches thought it was just a mental thing and sent him back down to Triple-A to get some more at-bats in. After three years of less than average production, Dowd decided he wasn't going to make it back to the big leagues and called it quits.

Jefferson County had a huge metropolitan area to draw talent from. They pulled all the best players from the school programs and the recreational leagues. It was the clash of the titans! We entered the tournament as the number one seed and the Cardinals were the number two. Both teams knew they were good and neither team was going to back down from the other one. We couldn't wait for the tournament to start! We were ready to dominate!

Day one got off to a great start. Our first game was at 10 a.m. on Saturday morning. We played a team from a smaller county from our bordering state. They had bussed in from six hours away. Their starting pitcher was solid, and the score was only 1-1 through three innings. We held them scoreless in the top of the fourth and then Metzler hit a three-run homer in the bottom of the fourth that put us up 4-1. Kevin added a two-run dinger in the seventh and we ended up winning the game 6-2.

Not that we were checking the scoreboard (we were), but we saw that the Cardinals had won their game 11-2. One step closer to a championship showdown. We also noticed that the Clark County Cougars had won their game as well. That meant that Coach James Churry and one of the only teams that we had fallen behind to the entire season would be advancing as well. Our second game was at 3 p.m. and would be followed by the Cardinals playing their game at 6 p.m.

We couldn't wait to start our second game. If we won that one, it would put us in the four-team championship bracket starting the next day. On the first day you played round robin against other teams, and the teams with the best record went on to play in a four-team tournament the next day. The teams would be seeded one through four and it was a single-elimination tournament.

We lined up for the National Anthem at 2:50 p.m. and were ready to play ball. We couldn't help noticing that some of the Cardinals players were in the stands watching our game. We were playing a team from our home state that we had seen in three tournaments that year. We had beaten them each time but knew never to get overconfident.

We got off to a quick start. Diaz forced a leadoff walk and proceeded to steal second on the second pitch of Gannon's at-bat. Not playing for a one-run lead, Coach Bowman was letting Gannon swing away. It paid off when Gannon stroked a base hit up the middle. With Diaz's speed and the fact that he got a great jump on the hit, there was not even a throw to home. The throw went to second instead. It was quickly 1-0 in the top of the first. We weren't done yet. Cooney followed with a base hit to left field and put our cleanup hitter up with a man on first and second with no outs.

With a 2-1 count, Metzler belted a deep ball to right field that we were sure was a home run. The right fielder had to make the catch at the wall, but it was enough to advance the runners to second and third. Kevin hit a rocket groundball to third base. Unfortunately, their third baseman was able to hold the runners and still throw Kevin out at first.

I was up with two outs and a man on second and third. On the first pitch, expecting a fastball, I placed a perfect hit down the right field line. It was a little outside so I had to reach for it. Call it a bloop single, but it was enough to score both Gannon and Cooney and we were up 3-0. Schreiber and Williams both cracked base hits and I was able to score on William's hit putting us up 4-0. It wasn't until Donnelly flied out to second that the top of the first was over. We had stranded two runners, but we sent a message by batting through the order in the top of the first.

We ended up winning that game 16-3. Every player had a great game. I went 3-for-5 with four RBI's. Even though we won big, Coach Bowman made sure we were never cocky. It was always a handshake and "good game" when we lined up to shake hands.

We didn't say a word to the Cardinals players as they passed us to take the field for the next game. We knew they saw the scoreboard and we let that say everything.

A few of us hung around for a while to watch some of the next game. The Cardinals were a solid team and played the game hard. They won their game 9-4 and were saving some of their better pitchers for tomorrow. James Churry and the Clark County Cougars made it through as well. There were going to be some tough matchups tomorrow. We would play the Butler County Bulldogs in the first game and the Clark County Cougars would play the Jefferson County Cardinals immediately following our game. The winners of those two games would then play in the championship. The Bulldogs had a good team. We had faced them a couple of times this season but always came away with a win. They were one of the only other teams besides the Cougars that we had fallen behind to this year.

I couldn't wait until tomorrow! I had trouble getting to sleep that night because I was so excited. I finally drifted off to sleep thinking about hitting doubles into the gap and making diving plays at second base.

We played the first game at 9 a.m. As usual, Kevin and I were the first two players there. We started playing catch. Just a light toss to get warmed up. As the other players started to arrive, our excitement started to build. Today would be the culmination of the entire season. I wanted to be a part of the championship lineage that the Bombers had.

The Bulldogs were a good team, and we would be facing their best pitcher, Stan Hollins. We had seen him earlier this summer in another tournament. He had held us to six scoreless innings in the last game that we played them. We finally came back from a 2-0 deficit in the seventh inning and won that game only 5-3.

As usual, Hollins had great stuff. He had a high-70s fastball and then could make you look silly with his 60-mph change-up. The two pitches looked exactly the same coming out of his hand. For only the fourth time all year we found ourselves behind again. Mostly because of the pitching of Hollins. The Bulldogs scored one run in the top of the second and added a solo home run in the top of the fourth. It was déjà vu all over again, as Yogi Berra once said. Behind Hollins' pitching, we found ourselves down 2-0 again. We never got frustrated or ever thought about giving up.

Hollins had a one-hitter going until we came up in the bottom of the fifth. Martinez led off with a base hit. He was starting for Williams in center field. Jason Mills, our starting pitcher for this game, was able to lay a perfect bunt down the third base line and advanced him to second base. Diaz was then able to get him to third on a groundball up the middle that their second baseman made a great play on. He robbed Diaz of a base hit and likely saved a run with a diving stop to his right. The throw barely beat the speedy Diaz to first base as Martinez rounded third. Gannon got us on the board with a rip down the third base line. It landed about two feet into fair territory and rolled all the way to the left field corner. He easily jogged into second with a stand-up double. We had cut the lead to 2-1.

It stayed that way for two more innings until we scored three runs in the bottom of the seventh. The Bulldogs made it interesting by answering back in the top of the eighth with a run. We then scored two more in the bottom half of the inning and won the game 6-3. We were going to be in the championship game! We all high-fived each other and were excited to be moving on.

We sat as a team to watch the next game. I have to admit, I kind of wanted the Cardinals to win. I had a ton of respect for James Churry's program, but we wanted a shot at the Cardinals.

Their game was the complete opposite of ours. Where ours was a defensive battle for the first several innings, their game started off with fireworks. I don't know if it was nerves but the starting pitchers on both sides struggled. The Cougars scored four runs on five hits in the top of the first. This included three singles and two doubles. Their fans, mostly friends and family, were going crazy in the stands. The Cougars players never felt more confident and were stoked to have such a good start against a talented team. Just in case they forgot who they were playing and wanted to get overconfident, the Cardinals quickly loaded the bases with no outs on a single followed by back-to-back walks. That brought up the Cardinals' cleanup hitter, Tommy DeMarco.

DeMarco was the Cardinals' best player. It was already rumored that Coach Dowd had spoken with the current baseball coach at the University of Missouri and that DeMarco only had to say the word and he would have a full-ride scholarship there. He was hitting .476 this year with 18 home runs and 47 RBI's. Keep in mind, this was over only 27 games! He was a pure hitter and could hit for power even to the opposite field.

CLINK! The entire stadium heard the unmistakable sound of the aluminum bat. DeMarco knew it as soon as he hit it and started his home run trot around the bases. He smiled toward the Cougars dugout as he jogged down the first base line. This infuriated Coach Churry but he kept his cool. He and Coach Bowman had similar philosophies, and the only retaliation would be runs on the scoreboard. (Many teams would have plugged De-Marco in the thigh during his next at-bat.) The Cougars left fielder took a few steps back but quickly pulled up and watched the ball go over his head. It was a monster homer clearing the left field wall by a good 20 feet. Just like that, the Cougars' lead was gone and it was tied 4-4.

This seemed to take some of the wind out of the Cougars' sails. They were never the same after that titan-ic home run. The Cardinals played with a level of confidence that could be intimidating to other teams, and it affected the Cougars that day. The Cardinals went on to wallop the Cougars 15-6.

The Cardinals were impressive, but we were not concerned. We wanted to play them. We were extremely confident because we had practiced hard and knew we would never let down one of our teammates.

It would all be decided at 7 p.m. This was the match-up we had been looking for all year. As Coach Bowman jokingly said earlier in the year, "May the best team win!"

LESSONS FROM THE CREED:

If you know you have trained hard, then you will want to compete and win. You will fear no opposition. You will be confident in your abilities because of your practice and training. You will never give up on yourself or your fellow teammates. You will always perform at your highest level and capacity, giving it everything you have.

CHAPTER 6

Readily will I display the intestinal fortitude required to fight on to the Ranger objective and complete the mission, though I be the lone survivor.

This was it! I couldn't wait for the first pitch. I had finally made it to the championship game at the Midwestern Fall Classic. This was the high point of the entire season. The last game until next year. This is where the hours of practice would pay off, and where the price of sweat, dirt, and sometimes blood would come to fruition. Though I had a ton of respect for the Cougars, I was glad we were playing the Cardinals. We wanted to beat the best team fair and square for the championship.

I had butterflies when we lined up along the third base line for the National Anthem. These were not nervous butterflies, but pure excitement. "The Star-Spangled Banner" ended and we heard the familiar "Play ball!" from the home plate umpire.

Coach Bowman pulled us together for a quick talk. "Men, this is something you will remember for the rest of your lives. This is the culmination of all the hard work and practice we have put in the entire year. I know I have been tough on you all, but that is because I see the enormous potential you have and want to bring out the best in you. I want everyone to give everything they have today. Leave it all on the field."

He then paused for a second, looked every one of us in the eye and said, "I love you guys."

I got a chill down my spine, as I had grown to love this team as well. It is amazing when you go through struggles how much of a bond can form between players. I would never quit on these guys. A few of us shouted, "Love you too, Coach!"

He then looked over and said, "Metzler." Metzler picked up the cue and said, "All right, gang, hands in the middle! Win on three!" We all shouted, "One, two, three, WIN!" and took the field for warm-ups. With the best record over the season, we were the home team and would get the last at-bat.

We had Donnelly on the mound—our best starting pitcher. He had pretty decent stuff going that day, and we were able to get the Cardinals out three up and down in the first inning.

The Cardinals had Carl Howe going. Howe wasn't much of a power pitcher, but he could paint the corners. He had a natural left-to-right cutter, which made him tough to hit. If you were a righty like me, his pitch would look outside and at the last moment would veer back to catch the outside corner of the plate for a strike.

Diaz grounded out to the second baseman on our first at-bat. Then Gannon drew a walk with one out and we had something going. Unfortunately, Cooney struck out and then Metzler flied out to left field and the first inning was over.

The top of the second started off a little shaky. Their first batter hit a groundball sharply down the left field line that I thought for sure was going to go through for a double. Our third baseman, Gannon, made a diving play to his right, and the ball bounced off the tip of his glove. He didn't field the ball cleanly but was able to knock it down and hold the runner at first base. The second batter drilled a sharp groundball right up the middle. Both Kevin and I pounced to try to get the angle on it. Once I realized it was going to the shortstop side of second, I immediately veered off to cover second base. Kevin made a spectacular dive and fielded the ball. Then, without any time to think, he flipped me the ball from his glove to get the force-out at second. It was an awesome defensive display and one of the best plays that Kevin made all year. I planted my right foot and turned to fire the ball to first base with the hope of turning two. It was not meant to be. The runner beat the throw by a couple of feet. Fortunately, we were able to get the next two batters out and get out of the inning with no damage.

Kevin led off the top of the second and I was on deck. Kevin smoked a line drive that unfortunately was right at the left fielder, who only had to take one step to his right to catch the ball. I got up and worked the count to 2-2 before I grounded out to the shortstop. He fielded it cleanly and was able to throw me out by a good ten feet. Schreiber popped out to the second baseman, and we were quickly out one, two, three. We could see that Howe was going to be tough to hit that day.

We were able to get two runners on in both the third and fourth innings. We just could not get anyone across the plate! It was a pitcher's duel and the score stayed 0-0 until the bottom of the fifth inning. This is when we finally made some noise. Our pitcher, Donnelly, led off the inning with a base hit. Diaz laid a perfect bunt down the third base line. Even though he was thrown out, he was able to advance Donnelly to second base. We finally got on the board when Gannon stroked a line drive up the middle, allowing Donnelly to score with no throw from center field. We were up 1-0.

Cooney followed with a base hit of his own, and we had men on first and second with only one out. We had a great chance to have a big inning! Metzler tallied our third hit in a row with a blooper right over the second baseman's head. Gannon scored from second and Cooney advanced to third on the base knock. We were up 2-0! We had first and third with only one out. Kevin came up next and lofted a high flyball to the third base side. It was in foul territory, but the third baseman still made the catch and was able to hold the runners.

It was now up to me to keep the inning alive. I worked the count to 3-2 before I was called out on strikes. Howe threw a cutter that looked a foot outside until it came back and caught the outside of the plate. "Strike three!" the ump called, and the inning was over. I was disappointed but knew Howe had beaten me fair and square. I was 0-for-3 for the day and really wanted to contribute to the team. The good news was that we had the lead 2-0 and had five innings in the books.

The Cardinals answered back in the top of the sixth with a run of their own. The score stayed 2-1 into the seventh inning, when I came up for my third at-bat. I was batting fourth that inning and got my chance to hit when Kevin took ball four for a two-out walk. I worked the count to 3-1. This is a hitter's count and I was looking for a fastball all the way. I hit the ball solid, but it was on the ground right at the first baseman. The pitcher charged over to first base, but the first baseman waved him off and I watched him step on the bag in front of me. I had no hits after four at-bats.

We brought in Mark Golden in the eighth to pitch the last two innings. He could throw some heat. He had been a relief pitcher all year long and had closed out many games. He pitched a gem in the top of the eighth, striking out the first two batters he faced. He then forced a groundball to Kevin, and we were three outs away from the championship! We even added an insurance run in the bottom half of the eighth inning when Schreiber blasted a leadoff home run. We were up 3 to 1 and excited to take the field for the top of the ninth!

It was the top of the ninth inning and we only needed three outs to have our championship. Golden struggled with the first hitter. The batter worked the count full before walking with a questionable call that I thought was right at the knees. The umpire thought otherwise and called out, "Low, ball four!" They had a leadoff walk. Down 3 to 1, this was no time to lay down a bunt. The second batter was swinging away. Golden quickly had the count one ball and two strikes and wasn't going to give him anything to hit. He got him out swinging with a nasty change-up that took a dive right in front of home plate.

We only needed two more outs! I got into action on the next batter. He hit a hard groundball to my left between me and Cooney. Since Cooney was holding the runner at first, there was a bigger gap than usual. I had to go hard to my left and made a diving stop to keep the ball from going through to the outfield. As I popped up quickly, I realized my only play was to first base. The runner was out by a couple of steps. We were one out away!

The Cardinals had DeMarco on deck so we wanted to end it right here. The next batter was the Cardinals third baseman. On a 2-2 count, he hit a lazy flyball to left center. It wasn't very deep, and Metzler, Diaz, and Kevin were all charging for it. He could not have placed the ball better if he had dropped it in between them. Diaz made a last-effort dive for the ball, but it fell for a base hit. The runner on second base was going on contact and made it to home without even having to slide. This made the score 3-2, but we were okay. We still only needed one out.

With a runner on first and two outs, DeMarco walked up to the plate. He looked confident and tried to stare Golden down for a second before stepping into the batter's box. Golden was not intimidated. He had been here many times before. The first pitch was low and outside. Ball one. Golden fired the next pitch right at the knees, just catching the inside corner. "Strike!" the umpire called out. The count was 1-1. Golden tried to go back inside again, but this time DeMarco was looking for it. He launched a deep drive right down the left field line. Metzler went running back and to his right but could only watch this one. Fortunately, the ball was hooking left. "Foul ball!" we heard the third base umpire call as he raised his hands while running down the left field line to get a closer look.

Whew! We had gotten away with one on that pitch. The count was 1-2. We were now only one strike away from a championship! Golden went low and outside again on the next pitch, hoping DeMarco would chase it. DeMarco wanted nothing to do with it and held his swing. Two balls, two strikes, and two outs.

I can still remember the sound of the aluminum bat. Golden went back inside, but this time he left a little too much hanging over the plate. DeMarco took a big hack and got all of it. CLINK! The entire stadium heard the ball connecting with the aluminum bat. Once again, Metzler could only look up, but this time the ball was well fair. Dang it! The Cardinals had taken the lead 4-3. Golden was a little shook up and we all gave him some encouragement. He got the next batter to ground out and we were down to our last at-bats.

We all got back to the dugout a little shaken, but no one was down. Suddenly, Kevin yelled out, "Guys, huddle up!" We all gathered around Kevin and even the coaches stayed back.

Kevin shouted, "Do we ever give up?!"

We all shouted, "NO!"

"Do we ever quit?!"

"NO!" we all shouted again.

"This is what we have worked for all year!" He paused, slowly looked at all of us right in the eye, and said in a quiet calm voice, "Let's go take it!" We were so fired up!

We had Jimmy Gannon leading off followed by the heart of the order. We couldn't ask for anyone better to jump-start the inning. Gannon worked the count to 3-1 and then he delivered. He laced a solid base hit between third and short to get the inning going. We had the tying run on base with no outs, and Cooney, Metzler, and Price coming up.

Cooney took a curveball in the dirt for ball one. The next pitch was a fastball that Cooney fouled about twenty rows directly behind home plate. He took the next pitch inside for ball two. He was thinking fastball all the way on the next pitch and guessed right. He hit a line drive on a rope, which we thought was over the shortstop's head. The Cardinals shortstop leaped as high as he could and caught the ball in the very tip of his glove. We call that a snow cone. The shortstop made a great play and definitely robbed Cooney of a base hit, but we were fine. We still had one on with one out and had our clean-up hitter Metzler coming up next.

Metzler swung at the first pitch and hit a towering flyball down the left field line. The left fielder drifted over to his right and made the catch just outside of foul territory. Gannon acted like he was going to tag up and took off right when the catch was made. Unfortunately, he held up once he saw that the throw to second was right on line.

That brought Kevin up with two outs and a man on first. "Let's go, Kevin! You got this!" I yelled as I walked up to the on-deck circle. I wasn't worried. Kevin and the rest of us had our backs up against the wall before and we had always delivered.

"Strike one!" yelled the ump as Kevin took the first pitch. It was an inside pitch that caught a couple inches of the plate. Kevin stepped out and looked down at the third base coach. The coach just clapped his hands and gave Kevin some words of encouragement. With two outs, there was nothing to do but swing away. The next pitch was a fastball that Kevin swung hard at. He was way out in front of it and the ball was grounded foul about six feet outside of third base. We were down to our last strike.

The next pitch was low and outside. The pitcher was trying to get him to chase one but Kevin was able to hold off. The count was one ball and two strikes. The next pitch was a change-up and Kevin was out in front of it. He hit the ball with a ton of topspin down the third base side. I have to admit that I cringed as this would not be a tough play for the third baseman. The third baseman came in four or five steps and fielded the ball cleanly. The only play was to first base as Gannon was going on contact since we had two outs. Kevin bolted out of the batter's box as fast as he could and sprinted down the first base line. The third baseman came forward and threw sidearm to first base.

I had a great seat standing in the on-deck circle. Kevin had average speed and it looked like the throw would beat him by a good five feet. My heart sank a little as I realized this would be the last out. *Pop!* I heard the ball hit the glove as Kevin was still sprinting down the line. Next thing I knew the first baseman bobbled the ball! There was so much spin from the throw that the ball actually fell to the ground. I couldn't help but flash back to the lesson that Coach Bowman had taught Kevin about never giving up on a play and running all the way through the bag. Kevin had never let up on running to first after that again. This was not only in games, but in practice as well. The first baseman reached down and picked up the ball just as Kevin's foot hit the bag. "Safe!" the first base ump shouted. I pointed at Kevin and pumped my fist at him.

Coach Dowd immediately came out of the dugout. He started arguing with the first base ump about the call. The Cardinals coach was arguing that the first baseman had picked up the ball before Kevin's foot hit the bag. It got heated for a while. The first base ump just kept saying "a tie goes to the runner; tie goes to the runner." Things cooled off after a few minutes and Coach Dowd walked back to the dugout.

It was up to me to keep the inning alive. We had a man on first and second and two outs. I had a pretty crappy day at the plate so far, going 0-for-4. This only made me think I was due. I stepped into the batter's box with my right hand up to the home plate umpire to call time and dug out my "go to" spot in the batter's box with my right cleat. Then I took a couple of practice swings and got into my stance. The first pitch was a low fastball right across my knees. "Strike one!" I heard the umpire shout. I thought it could have been called a strike, but it wasn't the pitch I wanted.

I stepped out of the batter's box and looked down at the third base coach. He gave me the "swing away" sign. I heard Kevin yell, "Let's go, Jones!" from first base. I stood back in the batter's box and did my normal routine; two practice swings and then set. The next pitch was another fastball but almost at chest level. "Ball!" the ump bellowed. I repeated the process by stepping out, looking at my coach, and then going through my at-bat routine. I got a curveball next. It looked juicy coming right down the middle of the plate, but I noticed the spin and was able to hold off. I was glad I did as the bottom dropped out and the ball almost hit home plate.

I had two balls and one strike for the newest, most important at-bat of my life. I got into my stance again, and the pitcher and I locked eyes. I wanted this at-bat! "Ball three!" I heard the umpire shout. It was another fastball but was about a foot outside. Three balls and one strike. I would definitely take a walk as this would get the tying run 90 feet from home plate.

The next pitch was a fastball right down the middle. I swung away and made good contact. The ball was down the first base line and flew right over the first baseman's head. This would do it! I knew we could at least score Gannon, and Kevin would likely be held up at third. As I was running down first, I noticed the ball had a lot of spin on it and was slicing toward the first base line. "Foul ball!" I heard the first base ump yell as he put both hands up in the air. Man, it was so close! I got my composure and started walking back to home plate. Kevin jogged over and gave me a high-five. "You got this!" he said as he jogged back to first.

I don't know how many times I had played this scenario in my head as a kid. Bottom of the ninth, two outs, and a full count. Only difference is this was not in my backyard or on our neighborhood street. This was real and was the most important at-bat of my young life. I grabbed my bat and took a couple of deep breaths to relax. I took a couple of practice swings and then settled back into my at-bat routine. After digging in and a couple of swings, I got set.

He came right after me. Another fastball in the strike zone.

CLINK! I hit this one solid as well. Another line drive in the exact same spot as the last one, right over the first baseman's head! I was sprinting down the first base line and could see the ball starting to slice foul again.

But this time it was different. It landed only six inches in fair territory and was headed toward the right field wall. I could see it take a nice bounce off the wall. It was hugging the wall as it had so much spin on it and was heading toward the right field corner. I could hear the crowd going crazy! I knew this would tie the game as Gannon had some speed and was going on contact with two outs. With Kevin's average speed, he would be held at third and we would have the winning run at third.

The right fielder got to the ball and was throwing it to the cutoff man as I was rounding first. Suddenly, the crowd got a little louder. Sure enough, I looked around toward third as I rounded first base and the third base coach was waving Kevin home! We would have a play at the plate to either win the game or go into extra innings! All I could do was watch at this point. Kevin hit the corner of the bag and was full speed heading toward home. The relay throw was clean, and the second baseman turned and threw a one-hopper right on line to home plate. Kevin went in headfirst and slid wide right of home base, hoping to sneak his left hand across the plate. The catcher fielded the ball cleanly and made a swiping tag right in front of the plate, but it was a fraction of a second too late. Kevin's hand slid right underneath it.

"Safe!" I heard the home plate umpire yell. I started jumping up and down and running toward Kevin. I only made it to about the pitcher's mound as I was mobbed by my teammates. We all hugged and were jumping up and down. I saw Kevin running toward us and he just jumped into the middle of everyone. With his momentum, we all ended up falling over laughing and yelling the whole time. My other teammates from the bench got there right after that and just piled on. We eventually recovered and there were hugs and high-fives all around. I found Coach Bowman and thanked him for all he had taught me and for a great season. He told me he was proud of me and never doubted me for a second. That meant the world to me. We were the Midwestern Fall Classic champions!

I still look back on that summer as one of the greatest summers of my life. I learned that if you want to achieve excellence and be a champion, you must push yourself harder than you ever thought you could. I learned that if you are a part of a team that wants to achieve great things, you have to grow as an individual and never let your teammates down. I also learned that if you go through challenges and struggles with someone, you will have a bond that others who haven't will never understand.

Go be a champion!

LESSONS FROM THE CREED:

Even if it comes down to only you, have the character and belief in yourself to never give up and do whatever it takes to get the job done.

RANGERS LEAD THE WAY!

AFTERWORD

I joined the U.S. Army the summer after I graduated high school and about a month after my eighteenth birthday. I remember my older brother, who was currently serving in the Army, told me to get everything in writing that the recruiter was promising me. With that in mind, I remember walking into the recruiting office and saying "Give me every school you've got! I want Airborne and Ranger!" I am sure they were licking their chops as I was an easy sell. My parents were not as happy that I chose the infantry as my military occupation versus learning a trade that would translate to a career outside of the military.

I am forever grateful of the life lessons learned and the pride I have by serving in the 2nd Ranger Battalion. The adage of "leave a boy, come back a man" was never truer for me. The leadership skills I was fortunate enough to observe and then duplicate have stuck with me my entire life. I was fortunate to serve under some great leaders in the Ranger Battalion. At age 19, I completed Ranger School and graduated in class 3-88. Rangers sarcastically called class 3 "class freeze" as it ran from mid-December through mid-February. This was the most challenging thing I had ever accomplished up to that point.

I will always be extremely honored to have served in such a distinguished and historical unit. I still use the Ranger Creed as my standard today in the pursuit of my own true potential. I carry an immense pride with me to this day that I was part of such an elite group of men.

RANGERS LEAD THE WAY!

Stephen today.

Stephen when he played tee ball.

Stephen in the Army Rangers.

ABOUT THE AUTHOR

Stephen M. Davis was born and raised in Marion, Indiana. He joined the Army a month after high school graduation and had the honor of serving in the 2nd Ranger Battalion, 75th Ranger Regiment. Serving in an anti-tank section, he earned his Expert Infantryman Badge, Pathfinder Badge and Ranger Tab. Mr. Davis and his wife now own an employee benefits company and enjoy having their son, daughter-in-law, and daughter all working in the family business. Now residing in Florida, they love enjoying the outdoors whether that be on the water, beach, or hiking.